HANDBOOK FOR WIRING GUIDE

How to choose type, electrical component to use and load calculation

Victor L. Kenneth

Table of Contents

CHAPTER ONE

INTRODUCTION TO HOME WIRING

In a House we use parallel connections which suggest that there will be separate change for each appliance.

Single section connection in home

This article is going to be very beneficial for you, if you desire to examine about how appliances, Motors, Generators, Light bulbs, circuit breakers, go with the flow switch, and Contactors are wired up. Single Phase wiring set up is the most frequent wiring in our residences and residential buildings. The Single Phase provide is 220Vac supply, which consists of two wires, one wire is Live and the different one is Neutral. These stay and Neutral wires come from the distribution transformer to the power meter and then at once linked to the predominant distribution board MDB in our homes. Single segment strength meter is linked with pole which receives 220V furnish from the step-down distribution transformer. The strength meter is then linked with the distribution box. The distribution field carries special circuit breakers which materials voltage to distinct rooms in a house.

Above is the easiest electrical wiring gadget for the Fans, Bulbs and outlets. The equal wiring device can be used for the Exhaust fans, TV, and different small load electrical family appliances.

CONNECTION OF SINGLE PHASE

Connect the stay and impartial wire from the electrical pole to the strength meter.

Connect the electricity meter with MCB circuit breaker

Finding the desirable measurement for electrical wiring installation

When we do residence wiring we first wants to comprehend which dimension of wire need to be used; We recognize that all conductors and cables have some quantity of resistance if we use small gauge wires for excessive load appliances, the wires will warmness up and motive brief circuit.

This resistance is immediately proportional to the size and

inversely proportional to the diameter of conductor i.e.

$R \propto L/A$

This equation suggests us that when the size of the wire will increase its resistance will amplify and when the place of the wire will increase its resistance will decrease.

A voltage drop happens in a conductor when present day flows thru it. Generally, voltage drop may additionally forget for small size of conductors due to the fact small size wire resistance is much less however in case of a decrease diameter and lengthy size conductors, we comprehend from the equation that resistance is immediately proportional to the size when the size of the conductor will increase it resistance will increase. We have to take into account the massive voltage drops for perfect wiring set up and future load management.

CONSIDERING THE WIRE TYPES

Aluminium and copper are two usually used wires. Copper has excessive in tensile strength, excessive conductivity, can be effortlessly soldered and is greater ductile. Although silver is the great conductor, its use is restricted due to the fact of its excessive cost.

Copper is extra high-priced than aluminum. Aluminum wire has about 60% of the conductivity of copper. It is used in high-voltage transmission traces and now and again in home and industrial wiring. Its use has expanded in current years.

CONSIDERING SIZE OF WIRES

Wire gauge is used to decide the proper measurement for a precise software due to the fact if use incorrect measurement of the wire it can motive injury to the equipment or the wiring system. Sizing of wire is finished by means of the AWG which skill American wire gauge system. It tells us the dimension of the wire and how tons present day it can safely handle. Common wire sizes are 10, 12 and 14 a greater wide variety corresponds to a smaller wire size, and influences the quantity of strength it can carry. For instance 10AWG wire has 2.588mm diameter, whereas a wire of gauge 20 has a diameter of 0.812 mm. So we can see that as the wide variety will increase the diameter decreases so these two are inversely related. A wire of 10 gauge can cope with a modern of 30 Amperes the place wire gauge of 20 can cope with 5 Ampere current. So as the gauge wide variety make bigger the quantity of cutting-edge that the wire can deal with decreases. The resistance per size for 10 gauge wire is 1 whilst the resistance per size for 20 gauge wire is 1 . This indicates that when the gauge wide variety is growing resistance per length is additionally increasing. For example, a low voltage lamp will require 18 gauge wire to draw 10 Amperes current, whilst carrier panels or subpanels require two gauge wire to draw one hundred Amperes current.

CHAPTER TWO

INTRODUCTION TO TWO WAY SWAP CONNECTION

Two way change connection is generally used in stairs. This kind of change is beneficial to use it in pinnacle and backside of stairs or one of a kind doorways in the room. The swap we usually use in our domestic has two terminals. While the change that is used for the two way change connection has three terminals

Common

Line 1

Line 2

The swap can be flipped in solely two approaches both up or down.

Major Components Required For This Connection

Two pieces of 2-way switches

Electric Bulb

Connection wires

Working Of Two Way Switches

Let believe that we are controlling a mild via two switches. Connect the stay wire with the frequent of swap 1. Connect the bulb one terminal with frequent of the 2d swap and different terminal of the bulb with the impartial wire. Now join the L1 of swap one to the L2 of the swap two Connect the L2 of change 1 with L1 of swap 2.

This work like special OR gate.

When the frequent of Switch one is related with L1 and frequent of change two is linked with L2 then the circuit will be completer and bulb will be flip on.

When the frequent of swap one is linked with L1 and frequent of swap two is linked with L1 the circuit will be damaged and bulb will be flip off.

When the frequent of swap one is related with L2 and frequent of change two is related with L2 the circuit will be damaged and bulb will be flip off.

When the frequent of change one is related with L2 and frequent of swap two is related with L1 the circuit will be whole and bulb will be flip on.

Single segment motor two way change wiring

There are conditions when you want to manage a single section motor Water Pump from two exclusive places two houses, two floors, etc. In a scenario like this the single section motor wiring is accomplished in such a way that if the motor is become on from one region it can't be became off from the different location. In this gadget we have created interlock machine such that if the motor is on from one domestic it can't be became ON from the different domestic till it is became OFF from the first home. The extraordinary benefit of this interlock device is that if there is trade in line and impartial no quick circuit will occur.

WORKING OF SINGLE SEGMENT MOTOR TWO WAY SWITCH

The impartial and stay wires that come from the most important electricity meter are related to the double pole MCB circuit breakers. The cause of the use of double pole MCB circuit breaker is that if we desire to alternate any appliance, laptop or load, or we want to restore some thing in the circuit, we can without difficulty flip off the grant via the MCB. Contactor will be used to flip ON and flip OFF the motor, it is used for the interlock gadget which is now not viable if we use a relay or a starter.

This contactor has three poles. The impartial and segment wires at the output, from each the contactors are common.

Contactor coils are labeled as A1 and A2. We will take a wire from the impartial of contactor and supply it to the A1. We will use contactor to our requirement for instance as provide is 220V we will use such contactor which make coil magnet from 220V. We will use two buttons for ON and OFF. Red button for OFF and inexperienced button for ON; the Red buttons used are of the kind usually closed and the inexperienced button used are of the kind commonly open. In these buttons we have each open and shut contacts. For purple button we will use shut contacts and for inexperienced button we will use open contacts.

Connect the section furnish to the shut contact of the crimson button. Connect the different terminal of crimson button to the open contact of inexperienced button via a wire. Now we will make maintain modern wire through connecting the button with the generally open contact of the contactor. Now we will create interlock gadget such that when first contact will be energized the other contact will no longer be energized. We will join the inexperienced button terminal with the generally closed of the 2d contactor. Connect the backside commonly closed of the 2nd contactor with the backside of first contactor typically closed terminal and additionally to the A2 contactor on the first contactor. Now join the backside usually closed of the first contactor with the A1 and A2 of the 2nd contactor.

When we will supply provide via inexperienced button it will come to the 2nd contactor commonly closed contact, if this contactor is de-energized then it will got here to the backside of usually closed contact which is in addition linked with the A1 and A2 coil of the first contactor which will make first contactor coil energized. Then in commonly open contact we have keep wire now internally usually open contact will come to be closed. This contactor will be final energized till we flip off the push button. By the use of this device brief circuit hassle will be finished. The equal process will be used for the 2d contactor.

CHAPTER THREE

ELECTRICAL WIRE CALCULATION FOR MOTORS AND OTHER DEVICES

Start Wire Size Calculating

To calculate the wire measurement for motor we have to understand the following values:

Load

Voltage

% Efficiency

Motor affectivity is commonly stated on identify plate of the motor.

Cable measurement calculation for single section motor:

Let think that:

Load = 1KW

Voltage = 230V

% Efficiency = 80%

We recognize that

$\cos\Phi$ is the strength issue which is equal to 0.8

$P=VI\cos\Phi\times\text{efficiency}$

Now we will calculate the current

$I=P/(V\times\cos\Phi\times\text{efficiency})$

$I=1000/(230\times0.8\times0.8)$

I=6.79A

Every single section motor has 2% cutting-edge drop.

Current drop = 6.79×0.02

Current drop = 0.1358A

Total current=6.79+0.1358=7 A

To calculate wire measurement we generally used two sorts of conductors copper and aluminium wire.

GUIDE FOR ALUMINIUM WIRE:

1 sqmm=1.5A

So for 7A the wire dimension will be 7/1.5=4.66 sqmm

For copper wire:

1 sqmm=2.5A

So for 7A the wire measurement will be 7/2.5=2.8 sqmm

Cable measurement calculation for three segment motor:

Let think that:

Load = 10KW

Voltage = 440V

% Efficiency = 80%

We be aware of that

$\cos \Phi$ is the energy component which is equal to 0.8

$P = \sqrt{3} \times$ VI $\cos \Phi \times$ efficiencyNow we will calculate the current

$I = P/(\sqrt{3} \times V \times \cos \Phi \times$ efficiency $)$

$I = 10000/(\sqrt{3} \times 440 \times 0.8 \times 0.8)$

$I = 20.5$ A

Every three section motor has 5% modern-day drop

Current drop = 20.5×0.05

Current drop = 1.025A

Total current = $20.5 + 1.025 = 21.52$ A

To calculate wire dimension we generally used two kinds of conductors copper and aluminium wire.

For Aluminium wire:

1 sqmm=1.5A

So for 21.52A the wire measurement will be 21.52/1.5=14.34 sqmm

For copper wire:

1 sqmm=2.5A

So for 21.52 A the wire dimension will be 21.52/2.5=8.608 sqmm

Single Phase Submersible Pump Starter

The submersible motor are one of the most generally used motors. For the Single segment Submersible Water Pump we will

want the following factors in the
Distribution box.

CALCULATE THE OVERLOAD PROTECTOR

This thermal over load protector
work when over contemporary
drift via it

 It consist of bimetallic strip when
over present day glide thru it
wreck the circuit.

MAKING USE OF DPST switch

DPST is double pole single throw
switch. This swap work like double
pole circuit breaker we can

additionally use double pole circuit breaker in location of it.

THE CARRING CAPACITY OF THE CAPACITOR

Capacitor is related with the motor as it is single section induction motor so it require capacitor ,at begin up it supply main electricity element to the motor.

CONNECTION OF SUBMERSIBLE MOTOR

The impartial will be related with the change pinnacle terminal from the backside terminal of the swap we will join capacitor and the different terminal of capacitor will be related with the motor. The line wire will be related with the swap different terminal at the top. From the backside terminal of change we will join overload protector the different terminal of overload protector will be linked with the motor. The different terminal of capacitor will be additionally related with the motor.

MAKING USE OF FLOAT SWAP COMPUTERIZED WATER LEVEL

Float swap is linked with motor so that it can automated flip on and off in accordance to the stage of water tank. The drift swap is set up in water tank, when the degree of water is lowered in the water tank, flow change provide sign to the motor to flip on and when the water tank is full the waft swap supply sign to the motor to flip off.

BASIC COMPONENTS REQUIRED

Float switch

Magnet contactor

Overload relay

Double pole MCB

Single pole MCB

The glide change consists of three wires. To take a look at the flow change we will set the Multi-meter on continuity it carries red, blue and black wires in which black is impartial wire. We will join the multi-meter one probe with the black wire and the different with the blue wire when the glide swap will be at backside it will act as closed circuit and supply sign to the motor to flip on. It supply beep on multi-meter which exhibit that it act as closed circuit. Now join the one probe of the multi-meter with the black and different with the purple wire when the drift change will pass upward it act like a closed circuit.

GUIDE TO CONNECT FLOAT SWITCH WORKING

We will join miniature circuit breaker (mcb) which mechanically switches off electrical circuit at some point of an strange circumstance of the community skill in overload situation as properly as a misguided situation with stay and impartial wire at input. Connect the output of the mcb with the contactor input. Magnetic Contactor is for lossy magnetic glide generated with modern-day in winding of such gadgets as transformer, throttles, magnetic cartridges filters and circuit. Output of contactor is linked with the motor. Connect the wire of the drift change with the impartial wire. Connect the L1 of the contactor with the A1. Connect the blue wire of the flow swap with the A2 of the contactor.

GUIDE TO AUTOMATIC CHANGEOVER SWAP FOR GENERATOR

A clever way to construct an automated switch change is by way of the usage of two contactors together. It is about two, electrically controlled, circuit breakers. The contactors are no longer allowed to shut simultaneously, however solely one at a time. The two contactors are joined collectively with a 'mechanical interlock' mechanism. It will now not permit to each relays of the contactors to be in a closed position. Connect the foremost furnish of the meter with circuit breaker input. Connect the MCB circuit breaker output with the contactor L1 and L3 terminals. Connect A2 of the contactor with neutral. Now join the section and impartial from the generator at circuit breaker input. Connect the circuit breaker output with the

2nd contactor at L1 and L3. Connect the backside usually closed NC of the backside with the A1 of the 2nd contactor. Connect the NC of the first contactor with the generator section line. Connect A1 of the first contactor with backside NC of the 2d contactor. This machine will shape a magnetic interlock. Now join the backside of each contactors from the place we can join our load.